BAMBOO

BAMBOO

Published by:
FormAsia Books Limited
706 Yu Yuet Lai Building
45 Wyndham Street
Central, Hong Kong
www.formasiabooks.com

Published 2004
ISBN 962-7283-80-0

Written by Trea Wiltshire
Designed by Format Limited
Editorial Assistance by Peter Moss
Misa Matsumoto – Kyoto, Japan
Production Assistance by Jenny Choi
FormAsia Marketing by Eliza Lee

Bamboo Products: Honeychurch Antiques, Hong Kong
Pages 4-5 & 162, courtesy Heidi Falter.

Printed in China by RR Donnelley
Colour separations by Sky Art Graphic Company Limited

Source of photographs:
FormAsia Books Archives
Kwan Kwong Chung
Camera Assistant: Sathish Gobinath

BAMBOO

Trea Wiltshire

FormAsia

平生豈意傲烟霞獨抱虛心澗寒葉枝立兩竿
搖動萬竿幾經風雨不撓斜
辛酉季秋月寫于桂林

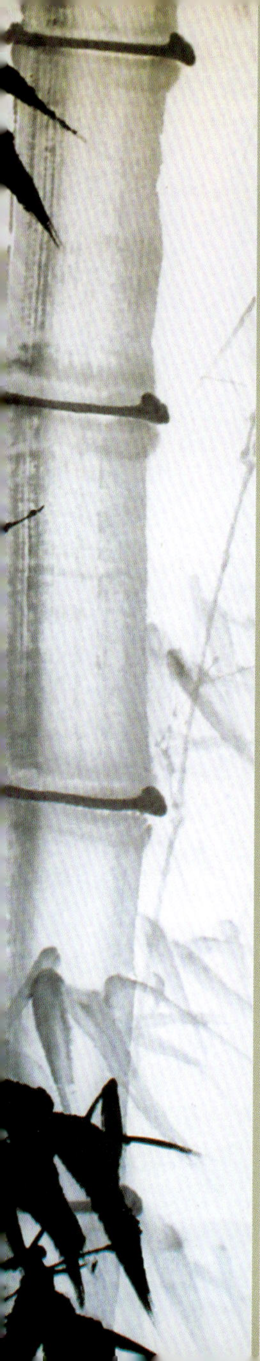

“ A meal should have meat, but a house must have bamboo. Without meat we become thin, without bamboo we lose serenity and culture itself. ”

A meandering,
bamboo-railed path
leads to Kinkaku Temple
(Golden Pavilion), Kyoto

Embraced by bamboo,
an alms box solicits pilgrims
visiting Ginkakuji Temple
(Silver Pavilion), Kyoto

浄

Water trickling through bamboo slakes the thirst and refreshes the spirits of worshippers at a Japanese shrine.

Screens of bamboo lend an air of mystery to the Ryoanji Zen Temple rock garden, Kyoto. The garden was laid out by landscape gardener Soumi in 1525.

Bamboo fenced garden, Koetsu-Ji Temple Kyoto. The garden is located on land donated in 1615 by Shogun Tokogawa Ieyasu, who lived in the vicinity with his family and relatives and founded a village for craftsmen.

Barricades of bamboo tower above the perimeter of Rakusai Bamboo Park, Kyoto. The park was established in 1981 by the Kyoto city government. Planted here are some 120 species of bamboo from all over Japan.

Extending over some 3.5 hectares, Rakusai Bamboo Park lies in the south-western suburbs of Kyoto.

Supplicant bamboo stems lean over a procession of monks mounting temple stairs at Inlay Lake, Myanmar.

Monks congregate in the shade on a bamboo dais at Mrauk-U, West Burma.

As serenely silent as their surroundings, novice nuns wend their way through a bamboo grove near Inlay Lake, Myanmar.

Bamboo backdrops grace pastoral idylls in Myanmar, where a farmer hoes his field and a water buffalo wallows in his element.

Waiting for high tide to speed a hot and sluggish journey to market, a boatman plunges into the Irrawady River, Myanmar, from his rafted convoy of fresh-cut bamboo.

An elegant frame for a patient pastime as youngsters cast their lines into the Ying Kwong River in South China.

A careful balance is required to ferry these oversized bamboo trunks down Myanmar's Irrawady River.

For centuries, bamboo-framed fish traps have been reaping Lake Inlay's piscatorial harvests.

POWER

The shade of a bamboo grove protects freshly potted ceramic bowls awaiting transportation down the Ying Kwong River.

Raised well clear of the highest flood levels on its bamboo piles, a thatched home overlooks the waterborne traffic of a klong in Thailand.

A dense mushroom cloud of bamboo leaves offers sanctuary from the sun to villagers in Uttar Pradesh, India, while a bullock cart conveys bamboo poles to a market in Bihar.

In India's Madya Pradesh, fresh bamboo shoots provide a meal-to-go for an elephant in workaday harness.

Parasols of bamboo boughs extend their shade over an evening bath in the outskirts of Rangoon, Myanmar.

Lining a village path in Chiang Mei, North Thailand, trunks of bamboo creak and sway in an evening breeze.

A champion Guang Ling rooster lays unchallenged claim to pride of place in the chicken run.

A little lad undid the rustic latch
as hand in hand your cottage we did gain,
where green limp tendrils at our cloaks did catch.
And dim bamboos o'erhung a shadowy lane.

Li Po (A.D. 701-762)
Village lane near Guangzhou

Frail, frail, lone-growing bamboo,
roots clasping the high hill's edge;
to join with my lord now in marriage,
a creeper clinging to the moss.

from Nineteen Old Poems of the Eastern Han Dynasty (25-220 A.C.)

Sheltered bamboo grove on the outskirts of Beijing.

This red archway, marking the main entrance to the Sun Yat-sen memorial park in Beijing, is flanked by bamboo to signify tenacity and endurance.

This sawmill in Guang Ling village, South China, works almost exclusively with bamboo.

Stacked sheaves of bamboo dry in the sun at Guang Ling, South China.

Bamboo wares still employ a significant proportion of China's population, and sawmills such as this one, in Sichuan Province, provide their raw materials.

Where once it was transported by barge and oxcart, most bamboo is now delivered in truckloads, as is the case in Dazu town, Sichuan Province.

The structure within may be Hong Kong's futuristic Space Museum, but the means of renovation is still supplied by time-honoured bamboo scaffolding.

The same material furnishes the framework for preservation of much older and more delicate structures, such as this stupa in Pagan, Myanmar.

Skeletal shapes spring up almost overnight in outlying areas of Hong Kong, to convey glad tidings that the annual itinerant opera troupe is about to return for another brief season.

Supplying the eternal matrix for the permutations of Hong Kong's changing skyline, bamboo scaffolding also provides access for its sign-painters.

When its initial outline is etched against a vaulting sky, bamboo scaffolding can seem as delicately woven as a gravity-defying mesh of cobweb.

Corralled behind a trellis of bamboo, this giant advertising hoarding appears to illustrate the classic Cole Porter cowboy ballad "Don't Fence Me In".

The bamboo organ in Las Pinas, Metro Manila has weathered earthquakes and typhoons for almost 200 years. It was built in 1816 by a Spanish monk who was also the parish priest of Las Pinas. Bamboo was the natural choice being both practical and aesthetically pleasing and emitting a truly unique, mellow tone.

Baskets woven of bamboo come in many shapes, sizes and forms, and the bamboo yokes that bear them are also employed to transport vegetables to market.

Below: Vapour rises from a backstreet stall where bamboo baskets are employed to steam the dim-sum.

Right: Rice cakes steamed in their bamboo tray, Dazu town, Sichuan Province.

意興隆通四海
油条大豆漿甜經濟

In Shanghai Street, Hong Kong, a solitary craftsman works under ever rising stacks of his own produce – the ubiquitous dim-sum baskets so vital to the restaurant trade.

山石書

Solemnly perusing the day's headlines, a venerable salesman awaits customers for his varied selection of calligraphic brushes, all fashioned of bamboo, set out on a bamboo table and attended by bamboo chairs.

No bloom she plucks to braid her hair,
but only sheaves of cypress spray.
In thin green sleeves
beneath the leaves
of tall bamboos she lingers there,
in winter's cold, at close of day.

Tu Fu (A.D. 712-770)

Zigzag bamboo *(Phyllostachys heterocyola mitf)*, one of many species found at Rakusai Bamboo Park.

No sound of the mid-day bell
enters this fastness
where blue mist rises
from bamboo groves;
down from a high peak
hangs a waterfall;
none know where he has gone, so sadly I rest,
with my back leaning
against a pine.

Li Po (A.D. 701-762)
Section of a bamboo grove, Jojakkoji Temple, Kyoto

Moonlight slants through
The vast bamboo grove:
A cuckoo cries

Haiku by Matsuo Basho (1644-1694)
Grove of yellow bamboo, Kyoto

BAMBOO

Trea Wiltshire

Since ancient times, bamboo has woven itself into the fabric of everyday life in China, touching countless events from birth to death.

A bamboo knife severed the umbilical cords of new born babies in remote mountain hamlets, and decades later bamboo provided the coffin and shoulder poles that carried the villager to his or her final resting-place. In much of Asia and South America, bamboo is still the 'poor man's wood', and those who work the land consider the graceful grass trees sheltering their homes to be far more than a fast-growing and incredibly versatile resource.

In Japan, bamboo implements are as integral to the tea ceremony as emerald pillars of bamboo are to the serenity of ornamental gardens. In Malaysia and Indonesia the anklung makes music, as do the haunting pan pipes of the high Andes, where multi-storey bamboo houses climb sheer hillsides. "Bamboo is our brother," say the Vietnamese, and in India its edible seeds have averted famine when crops have failed. And while grass trees belonging to the family Bambusaceae are native to every continent except Europe and Antarctica and boast more than a thousand species, bamboo is still most associated with China where it has become an important element of the rich and ancient culture that has embraced it for thousands of years.

China's long history is a cavalcade of imperial and revolutionary triumphs, of natural disasters and civil turmoil, with bamboo the ever present witness. It was called into action – as spear, bow, shield, barricade and gun barrel – in conflicts that won China an empire stretching from Korea to Turkestan. It was the great survivor that proved resilient to destructive typhoons, the blanketing snows of bitter winters, and failed or drenching monsoons. This beautiful yet supremely tough plant was the only living thing left standing in the heart of war-razed Hiroshima.

In China the primitive grass tree that is related to wheat and corn, barley and sugar cane, became the great provider. When warlords or civil strife scattered towns and villages, fleeing families searched for a grove of bamboo – the ubiquitous resource that flourished from tropical coastline to mist-wreathed mountain slopes, and provided virtually everything they needed to make a fresh start.

Bamboo's straight stems or 'culms' were easily cut and assembled as walls, and could be steamed and shaped to make furniture. The plant provided thatch for roofs, paper for windows, mats for floors, shavings for pillows, and the raw material for any number of household utensils. With their joints or "nodes" intact, the cylindrical culms could be cut horizontally or vertically and used as containers for water, honey or rice.

All this could be achieved swiftly and at

minimal cost, and every implement shaped from bamboo – from wine ladle to waterwheel – reflected the beauty and simplicity of its source.

Cut or split culms were quickly fashioned into an ingenious array of tools with which to till the earth, carry water to seedlings, and plant the staple that ensured a family's rice bowls were not empty for long. By cutting out interior joints, long culms became irrigation pipes feeding water from the nearest river to newly planted terraces, or were split and woven to form the mat-and-batten sails of a windmill.

Every year bamboo sent up shoots that the bare-footed farmer could feel pressing through the earth. If he wished to eat the tender shoots, he scooped a mound of soil around each to shield it from the sunlight that would toughen it – until he was ready to harvest this tasty delicacy. Herbalists also gathered the silicious tabasheer found in bamboo's joints and used it to treat a range of ailments from seasonal coughs to a decline in libido.

No matter how much bamboo a farmer

used, he knew that the grove would quickly regenerate, for bamboo is the world's fastest growing plant. It reaches its full height within months, and is ready to harvest within three or four years – far sooner than trees. And it is taller, for bamboos range from slender ornamental miniatures to forest giants rising more than 100 feet, and measuring almost a foot in diameter.

As night winds stirred the fronds above their homes, peasant families gathered to weave thin strips of bamboo into conical sun hats lined with sheaths of rainproof leaves. In skilled hands, the strips also became the ribs of a folding fan, a painted parasol, an ornamental cage to imprison a fighting cricket or house a songbird – or, of course, baskets. Conical baskets corralled protesting pigs and housed poultry and pigeons, while other varieties held rice and transported goods to market in carts or wheelbarrows also made from bamboo.

Its tensile strength was such that, lashed together, the culms were strong enough to span a river and bear the trundling carts and the farmers who in turn shared the weight of their baskets through bamboo shoulder poles. In making the bamboo cables of suspension bridges, farmers followed the example of their forefathers who pioneered such engineering feats in China as early as the fourth century.

Rivers, once the highways of China, were sometimes the only means of transporting produce from remote terraces to market towns. Indeed the rivers also carried vast rafts of newly cut bamboo, scoured with river sand to reveal its varnished, unblemished skin, then floated downriver to cities or ports.

The farmer might make his own bamboo raft – Asia's most ancient vessel – to punt produce to its destination, or his sons might build a small wooden boat, its sail fashioned from woven strips of bamboo and strengthened with bamboo ribs. The sails of this modest craft would be flat-topped, narrow and tall enough to catch the breeze above the riverbank, whereas those of the weather-beaten trading junks were shaped like butterfly wings. These sails were light, flexible, and easily furled when gathering clouds warned of an advancing typhoon.

The watertight bulkheads of Chinese junks were themselves inspired by the nodes that segmented a bamboo culm, and Western shipbuilders later adopted this feature that enabled damaged vessels to remain afloat by isolating compartments that were shipping water. Brother bamboo was also put to work in the Yangtze's churning gorges, where 300 harnessed workers, straining on narrow ledges high above the river, pulled vessels through the gorges – against the current – using bamboo cables. The cables, whose tensile strength rivaled that of steel wires, were made from the outer periphery of bamboo skin, a material rich in silica. When the river craft and trading junks reached China's market towns, the farmers and sailors again found bamboo everywhere – in the stalls that were quickly erected; the furniture of the local teahouse, the long-stemmed pipes of its patrons, and the tiny tools of the resident ear cleaner. In the heart of the town, they might see the mushrooming of an instant opera theatre that would soon be clothed with woven matting by nimble men moving across the bamboo scaffolding.

While farmer and townsmen alike put bamboo to work in countless practical ways, in market towns and cities it assumed additional roles, becoming the inspiration of scholars, philosophers and artists. In a quiet alcove, the flourishing gestures of a pavement calligraphist commanded attention, some of his brushes made entirely of bamboo, the fibre sharpened at the tips to deliver strong, incisive strokes. Others, held in a bamboo container, included some with plump stems of bamboo, used to make the bold, swelling movements that could define a subject with a single fluid stroke; while those slender as a reed seemed to take flight above the expanse of white paper, spinning lines as fine as silk.

Calligraphy and painting are closely allied, and both celebrate bamboo's unique status in Chinese art. Calligraphists write couplets in praise of bamboo's enduring character, while the artist endeavours to capture the grace of its plumes rising from valley or mountainside. Artists have always been drawn by bamboo's changing nature – by the restless interplay of light on leaves that gives the plant its intense vitality, and by the jade green light of deep-shaded groves that calls forth the most poetic sentiments.

Calligraphists, artists, poets and philosophers employ bamboo as a metaphor for much-admired human virtues. Its resilience brings to mind humankind's ability to bend before the winds of misfortune and remain resolute. Being evergreen, both pine and bamboo are familiar symbols of longevity, and of friends who remain true in adversity. Bamboo, plum and pine form a trilogy of plants with spiritual status that are revered as The Three Friends.

The most celebrated painters of bamboo are admired for capturing the essence of the plant and imbuing their work with ch'i, the universal life force of Tao. "To paint bamboo, carry the whole matter first in your mind. Then seize the brush, concentrate your attention, go straight ahead and follow what you see as quickly as the hawk swoops...", wrote a 13th century artist. "If you hesitate for a single moment, it is gone!.. Every leaf, curved or straight, beaten by rain or fluttering in the wind, has its own appearance. If they are all made the same, you've just blackened the silk".

Music also owes much to bamboo and when Chinese opera musicians warm up in a matshed theatre, most of their simple instruments are fashioned from culms of different dimensions: flutes and drum sticks, trumpets and fiddles.

Indeed, bamboo provided the world with some of its earliest musical instruments, and the plant's wide distribution ensured that it fathered distinctive forms of musical expression ranging from the anklung of the Indonesian archipelago to Japan's shakuhachi.

Bamboo groves make their own music: the gentle susurrus of leaf on leaf and the percussion of culm upon culm is soothing and unforgettable. The friction of culms sometimes creates a weakness in the plant's skin that a bird or burrowing insect might extend, and when a culm is pierced several times, it becomes nature's own instrument. In Malacca, a 19th century traveler described a forty-foot bamboo with man-made perforations, "through which the wind is permitted to sigh in a most charming manner". Locals called it the "plaintive bamboo" and its repertoire ranged from the soft and liquid notes of a flute to the deep, full tones of an organ.

The same traveler, intrigued by the different bamboos encountered in Asia, noted that its unpredictable flowering perplexed botanists. Some scientists observed species over decades without ever witnessing such an event, eventually discovering that few bamboos flower and shed seeds annually. In most species, inflorescence occurs at intervals ranging from 30 to an astonishing 120 years (in the case of Japan's most common bamboo, the madake). When this happens, all the plants of a species in a given locality "flower gregariously and die down".

This dramatic floral swan song that sees genetically-linked plants die at the same time – even those transplanted across the world – is one of bamboo's most distinctive features. Seeds shed will germinate in the leaf litter on the ground beneath a grove, but it may be a dozen years before a new grove reaches maturity. For people deriving an income from cutting culms or harvesting shoots the dramatic demise of a forest or grove can be a devastating occurrence. However, the ease with which bamboo is grown means that those dependent on the "poor man's wood" are seldom left bereft.

Culms grow from the tangled underground network of a rhizome that either throws out buds that become shoots, or spreads beneath the soil itself, sending out rootlets and buds. Given the right conditions – well-drained soil and humid tropical warmth – bamboo has been likened to a 'walking forest', so rapidly does it grow.

The rhizome stores nutrients and moisture gathered by its roots and leaves to nurture new shoots. To the Chinese, this interdependence is symbolic of a family struggling to survive. On achieving its desired height, the 'mother' culm takes no further sustenance, so the shoots can be nurtured. The self-sufficient grove also benefits from its leaf fall, which forms a moist blanket of mulch ideal for promoting growth and, later, for germinating seeds.

While two Persian monks had to smuggle the jealously guarded secrets of silk out of China (by secreting silkworm eggs in their bamboo canes), China willingly shared the bounty of bamboo with the wider world. By the 1860s bamboo was selling as a horticultural prop in London's Covent Garden, and when Europeans acquired a taste for China teas, they also coveted chinoiserie, ranging from lacquered bamboo furniture to paintings of the plant most associated with the exotic distant landscape.

In the 16th century, Christopher Columbus had admired "canes as thick as a man's thigh..." on his historic voyage, but it was not until the Victorian era that adventurous travelers fully grasped the amazing qualities of bamboo. Its phenomenal growth immediately excited the interest of botanists in herbariums around the world. In Calcutta they documented a culm of Dendrocalamus giganteus that grew more than 25 feet in a month, in Kiangsi Province in China it was noted that the extraordinary grass tree could grow one foot in 24 hours. And there was much botanical discourse about the differing characteristics of Bambusaceae encountered in locations as diverse as the jungles of New Guinea and the mountains of Abyssinia.

Entrepreneurs were soon examining the potential of the plant, and when Thomas Edison invented the first incandescent lamp in the 1880s, he tested a vast array of materials before opting for filaments of finely sliced bamboo from Kyoto – which were soon illuminating cities across the world.

Bamboo fibres were applied in a host of different ways in the West – from use as superfine gramophone needles to providing the strong but light material for a dirigible that a Brazilian aviator flew several kilometres (to the Eiffel Tower) in 1901. A less auspicious flight was Amelia Earhart's 1937 attempt to circumnavigate the world. For the grueling Pacific leg, the aviatrix emptied her plane of almost everything (from parachutes to a lucky talisman) bar an almost weightless bamboo fishing pole used to pass notes to her navigator.

Researchers around the world are only now appreciating bamboo's full industrial, agricultural, and environmental potential. It has already proved its worth in reinforcing concrete, as plywood and flooring, and as a readily available low-cost anti-seismic building material. It is also widely used in Asia to replace wood in paper manufacturing, and has the potential to serve as an alternative fuel.

Ecologically, bamboo is proving invaluable in landscapes from South America to China, strengthening the banks of flooding rivers, providing windbreaks against hurricanes, and binding soil to prevent landslides.

As forests decline around the world, bamboo – the 'walking forest' – is in wide use as a means of restoring diversity to landscapes. As cities expand, its graceful fronds are increasingly being called on to soften facades, muffle traffic noise, bring privacy to miniature gardens – and infuse the air with oxygen.

As the world begins to acknowledge that bamboo may well be its most universally useful plant, China's bamboo sector is already valued in billions of dollars, and the nation is in the vanguard of research that may well reveal additional roles for the grass tree. And while it may be largely eclipsed as a contemporary building material in China's rapidly-growing cities, in remote villages it has lost none of its usefulness – nor its mystique.

Villagers have always acknowledged bamboo's dual roles: as provider and protector. They know that each year a single hectare will yield enough material to build several houses, and that the whispering fronds that still protect their homes can be put to work in dozens of ways. And in hillside temples wreathed in incense, they still shake the chim sticks – those slivers of bamboo inscribed with characters – that tell of the arrival of a long-awaited grandchild, a generous harvest, and the good fortune that surely lies ahead.

Much may have changed in the 'Bamboo Kingdom' of the 21st century, but the words of a poet penned some 800 years ago still resonate: "A meal should have meat, but a house must have bamboo. Without meat we become thin, without bamboo we lose serenity and culture itself".

Bamboo Species from around the world

Green:	Most commonly cultivated in Japan. Used primarily for artifacts.
Green/gold:	Most widely cultivated. Used in making paper.
Green striped:	Found primarily in Japan. Ornamental uses. Turns yellow with age.
Golden:	Chinese. Used in the construction industry as scaffolding.

Mottled:	From Sri Lanka. Stains can be simulated by scorching the culms.
Tortoiseshell:	Resembling turtles. Can only be propogated from rhizome cuttings.
Black:	Native of China. Used primarily in decoration.
Giant:	Native of Burma. Extensively used for lumber, utensils and water vessels.

Tsuitati, a screen divider of latticed bamboo. Late Meiji Period c.1910. Japanese.

Bamboo stand for the display of Obi – a broad sash worn with Japanese kimono. Showa Period c.1930. Japanese.

Bamboo low table. Late Meiji Period. c.1910. Japanese.

Kitchen cabinet,
early 20th century.
Sichuan Province, China.

Kitchen cabinets, early 20th century.
Sichuan Province, China.

Above: Three-tiered kitchen cabinet.
Livestock was commonly stored in the lower two compartments.

Kitchen cabinet. Ventilation slats in lower compartments permit air circulation for livestock. Fujian Province, China.

Kitchen cabinet
with three separated
storage areas.
Decorated with
folk medallions.
Guangdong Province,
China.

Stroller and high chair.
Zhejiang Province, China.

Garden chair fashioned
of broad bamboo slats.
Rangoon, Myanmar.

Country chairs of varying size and design. Zhejiang Province, China.

Stained bamboo side-chair,
early 20th century,
Sichuan Province, China.

Pillow of woven rattan with
inner bamboo support.
Concave indentation. Chinese.

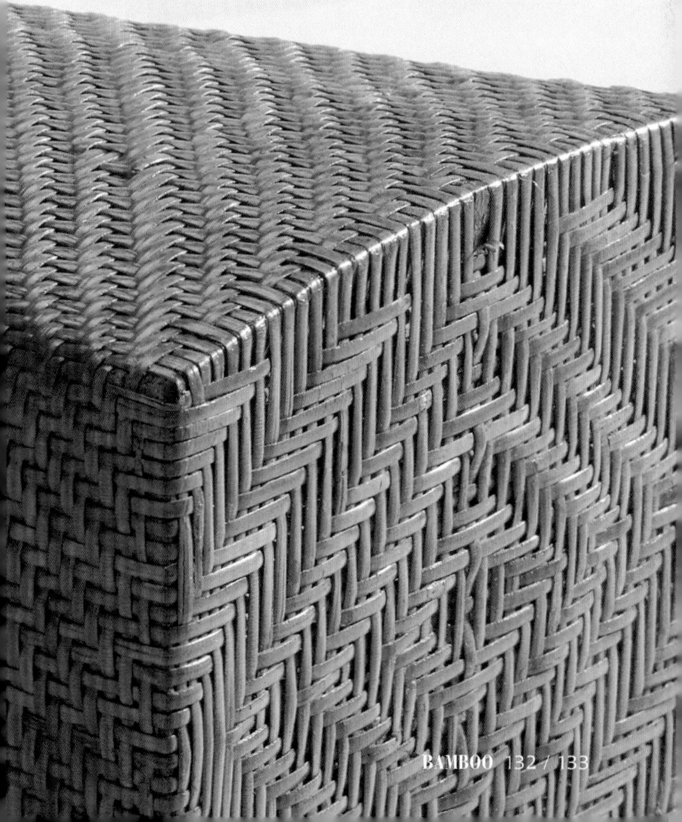

Hexagonal bamboo bird cage. Typical of Beijing area, for keeping the red-billed leiothrix. Hebei Province, China.

Cage with brass cup and latch to house the lark, favoured by bird-lovers of Beijing area, Hebei Province, China.

Cage with
stopper, woven
of bamboo
for retaining
chirping crickets.
South China.

'Mother and child chair', when stood on its side it turns into penned baby's chair with table top of slatted bamboo. South China.

Shellfish or fish trap woven of bamboo, with narrowing tunnel and spokes to prevent catch from escaping. Andaman coast – Thailand.

Left: Garden rake of bamboo.

Right: Straw, rattan and bamboo household sweeper. Thailand, Vietnam, Indo-China.

Pair of *renmain* 'human form' bamboo sections cultivated to achieve desired shape for decorative purposes. Late Meiji Period. c.1910, Japanese.

Brush pot of
lacquered bamboo.
Height 19cm. Hubei
Province, China

Zitan (purple sandalwood) brush hanger carved with dragon and cloud motifs displays an array of bamboo brushes. Early 19th century, Chinese.

大蓬头
畢
浙江湖筆

Brushes of goat and rabbit hair set in bamboo, used for either painting or calligraphy. Yiangzu Province, China.

Brush rest carved of boxwood to imitate bamboo – the whimsical fancy of some calligrapher. 18th century, China.

Huang hua-li brush holder with a variety of calligraphy and painting brushes set in bamboo. Fujian Province, China.

Brush holder of bamboo with two brushes, one of bamboo, the other of jade. Shandong Province, China.

Chinese tea leaf
scoop carved of
old bamboo.
19th century.
Shaanxi Province,
China.

Below: Carved ivory to imitate sliced bamboo – revealing the cocoon of an insect.

Right: Carved bamboo toggle.
Late 18th century. Chinese.

Two-tiered sweetmeat bamboo basket. c.1930s, Zhejiang Province, China.

丙寅年置

Two-tiered oblong
sweetmeat bamboo
basket, c.1930s
Zhejiang Province, China.

Painted sweetmeat basket with loop to suspend from bamboo shoulder pole. Swatow, Guangdong Province, China.

Ikebana basket. The arched handle made of grapevine root. Taisho period 1920. Japanese.

Ikebana basket of woven bamboo slats. The arched handle (previous page), composed of grapevine root. Meiji period. Japan.

Bamboo Ikebana basket.
Taisho period. 1920. Japanese.

Tiered bamboo lunch boxes of varying lacquer finishes with brass handles, early 19th century. Swatow, Guangdong Province, China.

Ginger grates of bamboo.
South China.

Above: Bamboo ladle used in the ritual of purification. Japanese

Right: Rattan noodle strainer with bamboo handle. Thailand.

Pairs of chopsticks for all occasions – including auspicious red for celebrations and freshly cut green bamboo chopsticks to mark a crossing of thresholds, on going to school or becoming employed for the first time.